W9-BQJ-488

ALMOST THE REAL THING

SIMULATION IN YOUR HIGH-TECH WORLD

The world's most famous simulator is Disney's Star Tours ride. See page 45. © *The Walt Disney Company*

ALMOST THE REAL THING

SIMULATION IN YOUR HIGH-TECH WORLD

by Gloria Skurzynski

with photographs chosen and arranged by the author

BRADBURY PRESS *New York*

MAXWELL MACMILLAN CANADA *Toronto*
MAXWELL MACMILLAN INTERNATIONAL
New York Oxford Singapore Sydney

Juv T57.62.S49 1991

For Ed, to celebrate forty years of the real thing

Acknowledgments

Again, heartfelt thanks to Ed, Lauren, and Jan.

The author is grateful to Neal Mayer of Evans and Sutherland Corporation and to Bob Judd of Los Alamos National Laboratory, whose contributions to this book were considerable and are greatly appreciated. Thanks also to Rick Adams of CAE-Link; Mike Gentry of NASA-Johnson; Naser Mostaghel of the University of Toledo (Ohio); Floreal Prieto of Arvin/Calspan; Don James and Steve Ellis of NASA-Ames; Chuck Hansen and John Fowler of Los Alamos National Laboratory; Dave Benman and Ann Lasko-Harvill of VPL Research, Inc.; Myriam Estany of Walt Disney Company; Keith Henry of NASA-Langley Research Center; Ted Yamada; Ted Alm; and David Nolan. Thanks also to Pixar; Hughes Simulation Systems, Inc.; Spectrum HoloByte; Mattel, Inc.; Nintendo of America, Inc.; Cray Research, Inc.; GPU Nuclear Communications Division; the National Geographic Society; Wright State University; U.S. Air Force Museum; Battelle Pacific Northwest Laboratories; and Flight Safety International. And very special thanks to Kristin Ferguson, pictured on page 59; to Tom Vavrin, on the back cover; and to Barbara Lalicki, my editor.

Copyright © 1991 by Gloria Skurzynski
All of the photographs in this book were chosen and arranged by the author.
All rights reserved. No part of this book may be reproduced or transmitted in any form or by any means, electronic or mechanical, including photocopying, recording, or by any information storage and retrieval system, without permission in writing from the Publisher.

Bradbury Press
Macmillan Publishing Company
866 Third Avenue
New York, NY 10022

Maxwell Macmillan Canada, Inc.
1200 Eglinton Avenue East
Suite 200
Don Mills, Ontario M3C 3N1

Macmillan Publishing Company is part of the Maxwell Communication Group of Companies.

First American edition
Printed and bound in Hong Kong
10 9 8 7 6 5 4 3 2 1

The text of this book is set in 13 point Sabon.
Book design by Christy Hale

Library of Congress Cataloging-in-Publication Data
Skurzynski, Gloria.
Almost the real thing : simulation in your high-tech world / Gloria Skurzynski : with illustrations chosen and arranged by the author. — 1st ed.
p. cm.
Summary: An introduction to physical and computer simulations, focusing on the use of wind tunnels, air and automotive safety testing, astronaut training, and current and future uses of images created by computer simulation.
ISBN 0-02-778072-4
1. Simulation methods—Juvenile literature. [1. Simulation methods.] I. Title.
T57.62.S49 1991
620'.001'13—dc20 91-238

CONTENTS

Engineers check this model on its way into a sixteen-foot tunnel that can simulate winds faster than the speed of sound. Simulations are imitations of things that exist in the real world. *Arvin/Calspan*

TESTING THE WIND
Chapter 1

Simulations are imitations of things that exist in the real world. Almost anything can be simulated—in images, in solid models you can touch, in sound, in motion, or in elements that you can feel, like the wind.

Just after the twentieth century began, Orville and Wilbur Wright predicted that people could fly. On a beach at Kitty Hawk, North Carolina, the brothers experimented with unpowered gliders. They wanted to try out different wing shapes, or airfoils, to find the one best suited to flight.

While Orville lay facedown in the center of a glider's lower wing, Wilbur and his helper ran along the beach, lifting the glider by the wingtips until it caught the breeze and flew, like a kite. They couldn't launch their flights unless the winds blew at about eighteen miles an hour, and they needed steady wind speeds to perform their airfoil experiments. But on the seacoast at Kitty Hawk, the wind and weather kept changing.

Using a bicycle they'd built in their shop, the Wright brothers tried to create enough wind to test the shapes of airplane wings. They clipped airfoil models to the top wheel, then pedaled to make it spin.
Wright State University Archives

Back at their bicycle shop in Dayton, Ohio, Orville and Wilbur tried balancing a spoked wheel horizontally above the front wheel of one of their bikes. Then they clamped airfoil models to the rim of that extra wheel. When they pedaled the bike, the airfoils spun in the breeze. The idea was a good one, but since it was hard to keep pumping the pedals at exactly the right speed, their test results weren't accurate.

Next the Wrights built a tunnel six feet long and sixteen inches square. Inside, a gasoline-driven fan blew twenty-five to thirty-five mile-per-hour winds across miniature wing models, while the brothers watched through a glass pane on top. Theirs wasn't the first wind tunnel ever built, or the first ever used to test airfoils. But the Wrights went about their experiments more carefully and thoroughly than anyone ever had before. In two months, after they'd tested 200 differently curved airfoils, Or-

The Wright brothers' original wind tunnel has been lost, but Orville Wright himself supervised the construction of this copy.
U.S. Air Force Museum

ville was able to say that he and Wilbur knew more about wing shapes, "a hundred times over," than all the earlier experimenters put together. Their tests led to the first motor-driven airplane flight at Kitty Hawk.

By creating a tunnel of moving air, the Wright brothers had *simulated an environment.* In small-scale experiments which they could control, they duplicated something that happens naturally in the real world.

Today, scaled-down models of airplanes or spaceships are placed inside wind tunnels for tests that *simulate* real flight. Instruments attached to a model measure the way the aircraft will react in the wind. Does it shake, shudder, or stall? Does it go out of control? Wind tunnel experiments let engineers try different designs without having to build and fly a whole new aircraft every time.

A skilled craftsman carves a model for testing in a wind tunnel. If its shape isn't identical to the real aircraft's, test results won't be accurate.
NASA Photo

The world's largest wind tunnel is at NASA-Ames in Mountain View, California. Aircraft are lifted on the yellow hoist and moved into the 80-by-120-foot wind tunnel behind it.
NASA Photo

The world's largest wind tunnel is at NASA's Ames Research Center in California. The two test sections of that tunnel are so huge that full-sized, real aircraft can be tested inside them, at wind speeds up to 350 miles per hour. Those speeds are useful for testing takeoff and landing maneuvers, but they're too slow for testing supersonic and hypersonic jets, which fly many times faster than the speed of sound.

In the wind tunnel.
NASA Photo

When all six drive fans are turned on full blast, 100 decibels of deep, rumbling noise rock the area around the wind tunnel. To get an idea of the fans' size, notice the people in front of fan number five.
NASA Photo

The National AeroSpace Plane, called NASP, will take off from an ordinary airport and accelerate into earth orbit. It will probably cost less to operate than a space shuttle. *Arvin/Calspan*

The National AeroSpace Plane, nicknamed NASP but also called the X-30, will fly at 18,000 miles per hour, fast enough to orbit the earth. No test tunnel now in existence can generate a wind that fast—it would destroy itself trying. Models of the NASP can only be tested in shock tunnels, where short bursts of compressed gases create shock waves that simulate *hypersonic* speeds—up to 25,000 miles per hour. In the milliseconds when the shock waves hit the models, photographs record the aerodynamic reactions of the model. *Aerodynamic* means the way an object responds to the force of moving air.

If all goes well, by the twenty-first century the NASP will take off from an ordinary airport, reach orbit, and carry supplies to the space station. It's also expected to carry paying passengers. In the NASP you'll be able to fly from Los Angeles to Tokyo in two hours. You may even be able to take a NASP flight into space, at five miles per second! Before that happens, the NASP will have to undergo many simulated flight tests.

Wind tunnels do more than test airplanes and space vehicles. At Battelle's Pacific Northwest Laboratory, in Richland, Washington, a stainless steel wind tunnel is used to test the effects of air pollution. Everywhere in the world, contaminants from power stations, factories, automobiles, pesticides, and smog fall on growing crops and cling to their leaves. When animals or people eat these plants, their health can be affected.

Since no wind tunnel can simulate orbital speeds of 18,000 miles per hour, a NASP model has to be tested in a hypersonic shock tunnel.
Arvin/Calspan

To test for health hazards, wind tunnels blow pollutants over living plants. A laser beam measures contamination during the test.
Battelle Pacific Northwest Laboratories

Since even a slight breeze will stir up and spread these pollutants, Battelle's researchers place living plants—and sometimes trees, animals, and even people—inside their wind tunnel for tests. They adjust light, temperature, and humidity to simulate conditions in the real world. Then they add pollutants.

As the test winds blow, laser beams measure the amount of contamination that sticks to the leaves. This lets scientists identify the level of airborne pollutants that will be harmful to human health, so their use can be regulated.

Wind tunnels are also used to test objects that speed over land or water, including race cars, speedboats, and athletes on bicycles or skis. At the Arvin/Calspan Advanced Technology Center in Cheektowaga, New York, members of the U.S. Olympic Downhill Ski Team practice their tuck positions in a low-speed wind tunnel. By reducing their bodies' wind resistance, they can shave tenths of seconds off their speeds. That may not sound like a lot, but for skiers racing downhill at eighty miles or more per hour, a tenth of a second can mean the difference between winning and losing.

In the past, speed skiers would strap their skis—with themselves on top of the skis—to the roofs of cars. As the cars roared along highways, the skiers practiced holding their poles, helmets, and themselves in positions that would minimize body surface. That was a wildly dangerous way to simulate conditions for a sport that's dangerous to begin with!

Testing in wind tunnels is much safer. In fact, one of the biggest advantages of simulation—no matter how it's used—is that it's safer than the real thing.

By streamlining body positions, skiers can shave tenths of seconds off race times—enough to go for the gold! *Arvin/Calspan*

A popular toy, the Slinky, gave Dr. Naser Mostaghel the idea for a new kind of base isolator to make buildings safer during earthquakes. In a laboratory test, the Slinky-like device is stressed sideways. *Naser Mostaghel*

SHAKE, SPLASH, CRASH *Chapter 2*

Have you ever played with a Slinky? Dr. Naser Mostaghel, a professor of civil engineering at the University of Toledo, in Ohio, says, "In 1978 my father-in-law gave a Slinky to my older daughter. That night I played with it, and the next day I bought twenty of them."

At that time, Dr. Mostaghel worked as a consultant for the Iranian Atomic Energy Organization. French engineers were building two nuclear power plants in southern Iran, an area where earthquakes are common. To protect the power plants from earthquake damage, the engineers built them with base isolators.

Base isolators are placed between the foundation of a building and the structure itself. During an earthquake they flex, which cuts down on the amount of energy that rises from the ground to shake the building. Different kinds of base isolators were already in use, but the Slinky gave Mostaghel a new idea.

A model of a five-story building, scaled to one-third size, is ready to be shaken on the table.
Naser Mostaghel

"I decided to separate the Slinky into flat rings, and to use a core of rubber at the center," he says. The new design, called R-FBI for Resilient-Friction Base Isolator, needed to be tested to see whether it would work. Since it wasn't practical to put it under a building and wait for an earthquake, Dr. Mostaghel took the device to the Earthquake Engineering Research Center at the University of California, Berkeley.

The earthquake simulator, or "shake table," located there is the first of its kind ever built and the largest in the United States. Engineers placed the Slinky-like base isolators on the twenty-by-twenty-foot shake table. On top of them they built a model of a five-story building constructed to one-third scale, which means one-third of real size and weight.

The shake table can be programmed to simulate any real earthquakes that have already happened somewhere in the world. Dr. Mostaghel picked six earthquakes of different magnitudes, ranging upward to the 1985 Mexico City earthquake that measured 8.1 on the Richter scale.

As an engineer counted down, "5-4-3-2-1!", the shake table began to shudder. It moves with a powerful rumble, backward and forward, sideways, up and down, so ominously that if you watch a simulated earthquake, you hope you'll never be caught in a real one! But the R-FBIs worked. They significantly reduced shaking in the building model, which showed they'd be useful in the real world.

Earthquakes aren't the only disasters simulated to research safety methods. Sometimes airplanes crash when they take off or land in severe storms; this kind of turbulent weather is simulated at NASA's Langley Research Center in Hampton, Virginia. Langley's heavy rain simulator, half a mile long, contains 1,590 nozzles. Water sprays through the nozzles to simulate rainfall of two, four, ten, thirty, or even forty inches per hour. Full-size airplane wing sections are thrust through the downpour at takeoff and landing speeds to see how well they'll perform in real storms.

Violent storms may be deadly for aircraft. At NASA-Langley in Virginia, the heavy-rain simulator duplicates rainstorms from squalls to monsoons. Clamped between disks, a section of airplane wing hurtles through a manmade deluge.
NASA Photo

If it's at all possible to get out alive from a plane crash, the right seat may make the difference. Seat and dummy will be bolted into a real airplane for a real crash in simulated conditions at NASA-Langley.
NASA Photo

Hoisted to the top of a frame that looks like a giant Lego, the plane gets cut loose. The dummy is in the cockpit.
NASA Photo

In other research simulations at Langley, real airplanes are dropped from a derrick high overhead to crash on the ground. This is done mainly to test seats and safety harnesses. In a low-impact airplane crash, well-designed seats can save lives.

Numbers on the grid behind the crash pad show how far the plane skids before coming to a stop.
NASA Photo

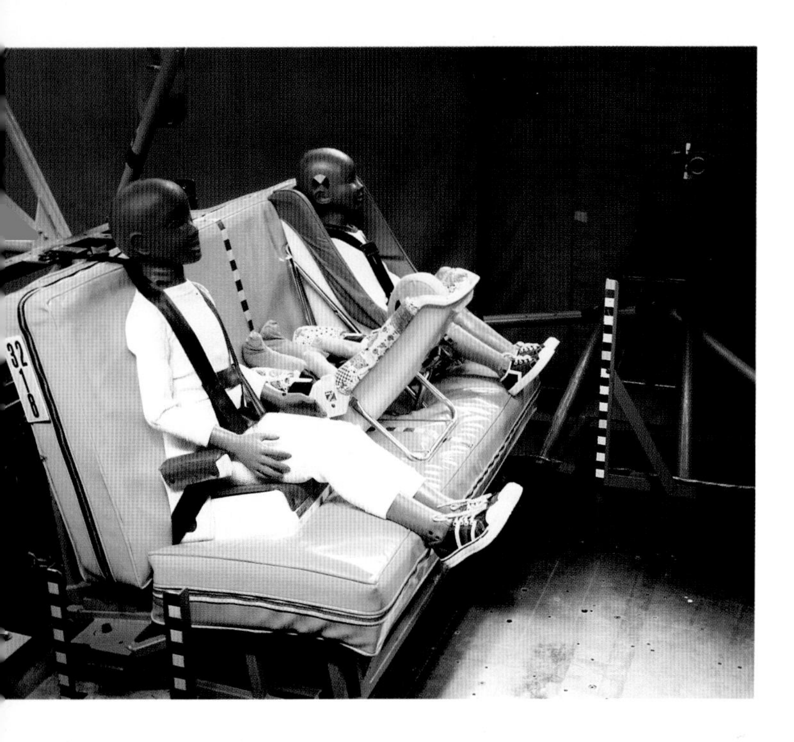

Electronically wired crash dummies are used to test whether automobile seat belts and child restraints meet federal standards.
Arvin/Calspan

For automobile safety, seat belts and child restraints are tested on specially wired, child-size crash dummies. They're strapped into seats that hurtle down tracks to smash against walls.

One way to learn how accurate simulations are is to compare them to real accidents. After every airplane crash, officials from the Federal Aviation Administration inspect all the wreckage, sifting through even the smallest pieces to discover such things as the angle of impact and the points where the plane broke apart. By looking at pieces of the crash, experts can tell whether earlier simulations were accurate. If a piece from the fuselage of the real crash is twisted exactly the same as one from a simulated crash, it's proof that the simulation test results were reliable. The same is true for automobile crashes. This is called *validation,* and it means testing the simulation against the real world.

As engineers monitor their consoles, car and dummy will shoot along on the high-acceleration sled. Simulations are valuable only if they're based on accurate data.
Arvin/Calspan

That's why earthquake simulators use measurements from real earthquakes, why plane crash simulations are compared to real crashes, and why a human engineer joined a crash dummy in a seventeen-mile-an-hour test crash—to see whether the dummy's body would react exactly the same as the man's. Everything checked—arms, hands, head, and body position.

You can learn a lot from a dummy!

An engineer undergoes a seventeen-mile-an-hour test crash to make sure the lifelike dummy reacts the way a human would. Everything checks—arms, hands, head, and body position. The dummy, though, doesn't seem to be suffering.
Arvin/Calspan

Physical simulations prepare astronauts for emergencies. In a seat built on slide rails, astronaut Mae Jemison gets ready to eject from a simulated mishap.
NASA Photo

GETTING READY *Chapter 3*

Astronauts' jobs are different from anyone else's, because they don't work on land or sea or in the air. They work in space.

In the almost-zero gravity of outer space, everything flies around if it's not tied down. Cheeks and eyebrows rise higher on faces, spines stretch out an extra inch or two, and shoulders hunch up because the arms don't weigh them down. Astronauts need to practice being weightless. But how is that possible in Earth's gravitational field?

There are two ways to simulate weightlessness without leaving Earth. One is to enter a neutral buoyancy pool: At NASA's Lyndon B. Johnson Space Center in Houston, Texas, the pool is called WETF (pronounced wet-if), for Weightless Environmental Training Facility.

Getting ready to go into the WETF is a big job. Several people have to help an astronaut into the bulky space suit. After all the lower layers are on and the space suit is tightly sealed, air is pumped inside to a pressure of about 3.5 pounds per square

In the Weightless Environment Training Facility (WETF) at Johnson Space Center in Houston, Texas, all attention is on two astronauts being lowered into the water. A mock-up of a shuttle's cargo bay is already submerged. Neutral buoyancy in pressurized suits underwater is much like the feel of near-zero gravity in Earth orbit.
NASA Photo

inch. If the astronauts were lowered into the pool just then, they'd float like inner tubes because the air-pressurized suits are lighter than water. To create neutral buoyancy—weightlessness—lead weights are added to the front and back of the suits, and to the arms and ankles, the way scuba divers wear weights to let them sink.

Next the astronauts are lifted by a hoist and lowered into the twenty-five-foot deep pool. Also in the pool are mock-ups of the payload bay or other hardware such as the Hubble space telescope. Underwater, the astronauts practice repairs they might have to make on equipment outside the spacecraft (inside it, they wouldn't need space suits).

Space suits are uncomfortable and tiring, even in zero gravity. Says astronaut Robert Stewart, "If the suit is fit properly, you're kind of wedged into it."

Astronaut Kathryn Sullivan adds, "You learn, through several runs in the water tank, a whole set of lessons that have to do with suddenly being a person of greater mass and volume." This training is essential because, Sullivan says, "The vacuum of outer space is a tough and unforgiving environment." Simulating outer-space conditions in the WETF lets astronauts practice for emergency situations ahead of time.

Space-suited astronauts Bruce McCandless (left) and Kathryn Sullivan (surrounded by divers) work on a mock-up of the Hubble space telescope. They're neutrally buoyant—they won't sink, nor can they swim to the top of the pool. In case of emergency, the scuba divers would bring them up.
NASA Photo

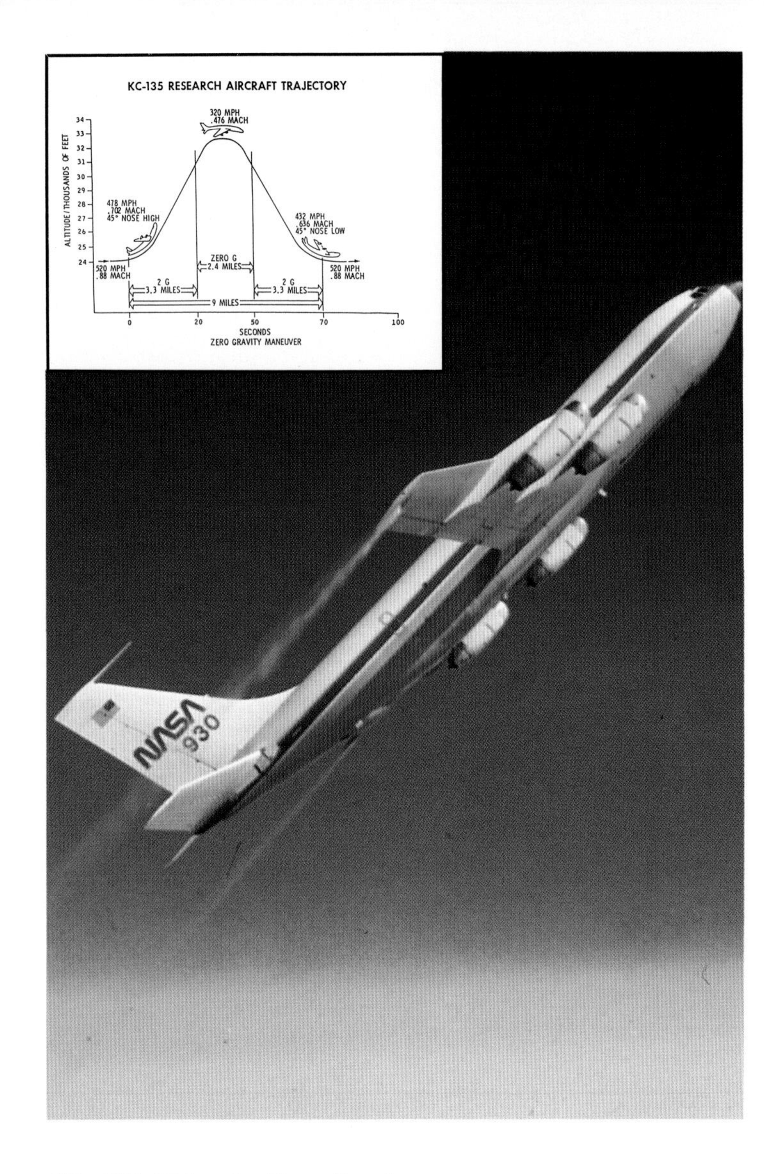

The KC-135—the Zero Gravity aircraft—climbs at a forty-five-degree angle. At the top of its ascent it will curve over and fly down at the same angle. During the curve, or parabola, everything inside will be weightless.
NASA Photo

The second way to simulate weightlessness is parabola flying. NASA's KC-135, a former Air Force refueling tanker, flies upward at a forty-five-degree angle to about 30,000 feet altitude. Then it curves around in a parabola and starts to descend at a forty-five-degree angle. For the thirty seconds or so that the airplane is at the top of the curve, everything inside it is weightless.

In these simulations the astronauts don't need space suits; they wear their regular clothes. The first couple of times people fly parabolas, about half of them get motion sickness: NASA provides white plastic airsick bags, and they're used!

Mostly, KC-135 flights are practice sessions for astronauts who will make space walks. Putting on space suits while weightless can be tricky! If the astronauts push too hard, they may fly across the cabin and bump into someone. This sets the other person spinning, and could cause injury.

During parabola flights, astronauts also try out experiments they'll perform in space, to learn how objects react in weightless conditions. In spite of the serious purpose of the flights, people in the cabin have a lot of fun floating—for thirty seconds per parabola. Then gravity sets in again.

The weightlessness they experience in these training flights isn't simulated: The trainees *really are* weightless during the parabolas. What's simulated is the environment of space. Inside

Each weightless period lasts only thirty seconds, but during each training flight, the KC-135 flies twenty to forty parabolas. Weightlessness is fun!
NASA Photo

As computers become more powerful and less expensive, they're used in all kinds of interactive training courses.
Hughes Simulation Systems, Inc.

the shuttle, in orbit, a pen accidentally dropped can drift away, and water spilled out of a glass floats in spheres all over the cabin.

When they're weightless, or ejecting from seats on slide rails, or opening a hatch underwater, astronauts are training with *physical* simulations—environments that can be touched and felt. Wind tunnels, shake tables, crash sleds, ejection seats on slide rails, and the WETF are all physical simulations.

More common than *physical* simulation programs are *computer* simulations. These are used to train people in all sorts of jobs.

You may have used a computer program in which you read a question on the screen, and you tapped an answer onto the keyboard. Usually they're simple answers—yes or no—and the computer tells you whether you're right or wrong. If wrong, think about it and try again. That's computer training, but it isn't simulation.

Computer simulation duplicates real job operations. For instance, in classes where Navy personnel learn about radar, computer programs simulate blips on a radar screen, and the sailors must identify those blips. Is it a ship? A submarine? Whales? The simulated blips are identical to the ones the radar operators will see on their screens when they begin active duty on the high seas.

Navy and maritime personnel train on computer-based radar simulators at a fraction of the cost of using actual equipment.
Arvin/Calspan

Computer-simulation training is most valuable in jobs where safety depends on fast, correct choices. At the Three Mile Island Nuclear Generating Station, operators learn their work in the state-of-the-art Replica Simulator of the plant's control room. The $18 million simulator is identical to the real control room—panel for panel, instrument for instrument, switch for switch.

If the trainees make a big mistake, such as forgetting to cover the reactor core with water during a plant shutdown, they go back and practice it over again. Since the mistake was made on the simulator, no disaster has happened, as it would have in the real world.

Simulation training is an excellent way to prevent disasters.

At the Three Mile Island Nuclear Generating Station, operators learn their jobs in an $18 million replica simulator.
GPU Nuclear Communications Division

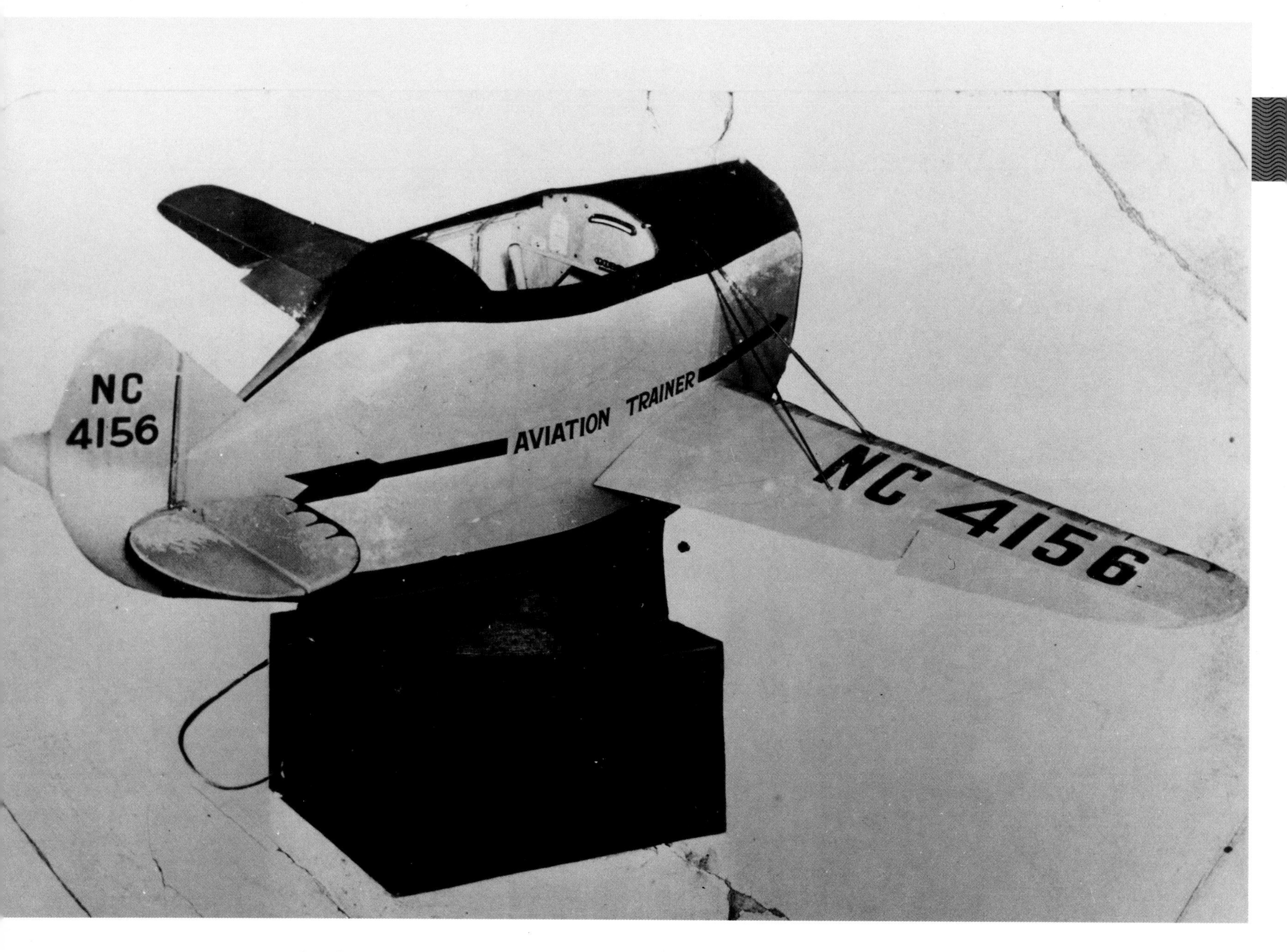

Edwin Link built his first flight trainer in 1929, in the basement of his father's player-piano and organ factory. *CAE-Link*

FLYING ON THE GROUND *Chapter 4*

World War Two. Flying a dangerous mission in his AT-6 single-engine plane, the pilot tensely watched the gauges: fuel level, power settings, wind drift. He was far off course and low on fuel. Below lay rough terrain; if the plane crashed in the dark, he'd never get out alive. The only way to survive was to bail out, right then, which he did: In one quick move he opened the cockpit's hood and hurled himself out of the airplane.

But his parachute never opened, because there was no airplane at all, only a mock-up of a cockpit standing solidly on the floor. Not till he fell the three-foot distance—and broke his ankle—did the embarrassed pilot remember he'd only been training in a Link aircraft simulator.

Edwin Link built the first flight simulator in 1929 in Binghamton, New York. Back then, flying lessons cost as much as fifty dollars an hour—more than most people could afford. Link's simulator, which sat on top of bellows that made it dive, climb, turn, and roll, shortened training time in real airplanes. Over

In the years between the *physical simulations* of the early Link trainers and today's *computer-imaged simulations,* model boards filled huge rooms. As the pilot handled the controls in a replica of a cockpit, a tiny TV camera moved over the board, sending pictures to a TV screen the pilot watched.
CAE-Link

the next decades Link added more instruments to the control panel and projected movies of sky and real terrain in front of the cockpit.

In the early 1970s, Evans and Sutherland Corporation, pioneers in the development of computer-generated images, produced the first visual images for flight simulators. Although they showed only points of light for night scenes, the simulated stars and city lights looked real enough to the pilots-in-training.

Today, both day scenes and night scenes are so realistic, they're like movies in a theater, except that they're interactive, like video games. When the pilot makes a move, the image changes in instant response.

An instructor at a computer can load every imaginable disaster into the flight simulation program: fires in the engine, wild turbulence, wind shear, iced wings, or even an engine falling off—something a pilot would have great difficulty recovering from. Still, the pilots work hard: sweating, groaning, and doing their best to land the plane safely, even though, in a tiny wedge of the mind, they know they're not in a real plane—it's all a simulation.

Nighttime visuals, seen through the windshield, are less expensive to make than daytime visuals.
Evans and Sutherland

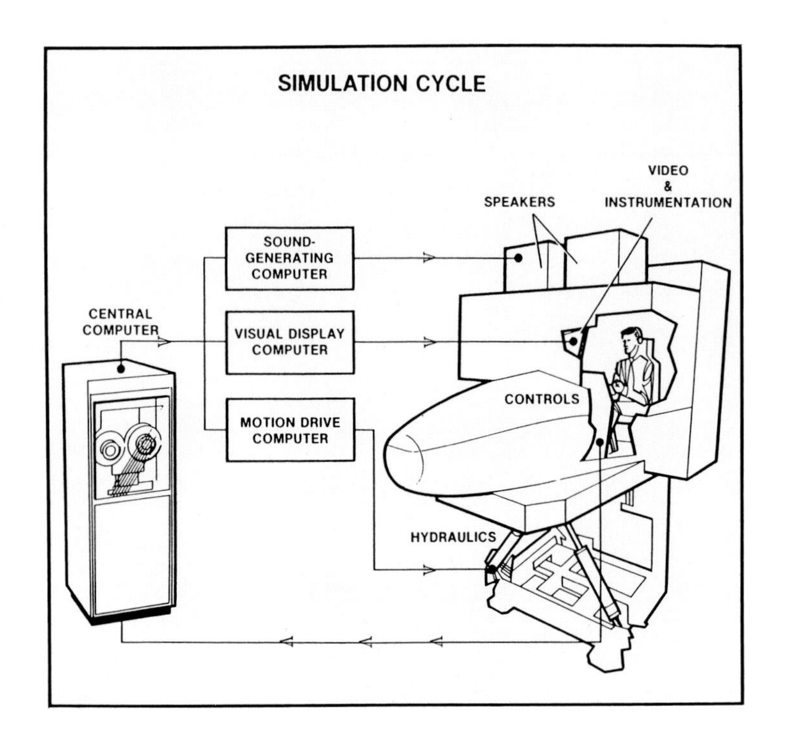

Signals travel from the computer to the simulator and back in only seventy milliseconds, so fast that no one notices the time lag. Audio, visual, and motion cues feel instantaneous, as though they're happening in "real time."
NASA diagram

The realism is created by electronic trickery. When the pilot moves the controls, input is sent to the computer. It responds with

Images: Through the window, you can see contrails from jet aircraft; fields with ground shadows; trees growing singly or in clumps; and the painted line on the landing strip, at first barely visible, and then growing sharper and wider until the front wheel touches down in perfect alignment.

Sound: An audio system pipes in the rush of the wind; the whine of the motors; the *whup, whup, whup* of helicopter blades; the throb of rubber tires meeting a concrete runway, bumpy or smooth.

Motion: Hydraulic legs move up, down, or sideways as the "plane" climbs or banks for a turn. Panels push against the pilot's back and seat to simulate the forces of gravity. G-suits, with balloon-like linings, expand and contract to squeeze the pilot's arms, legs, and chest, just as gravity does during supersonic dives.

The visuals, or computer-generated images, may be projected onto flat screens, or they may be projected inside a dome large enough to hold the cockpit. Since dome images have to cover such a wide area, they're sometimes dim and a bit fuzzy.

Wrap-around visual images are projected inside domes. *Evans and Sutherland*

According to Richard Adams of CAE-Link, "A simulator's key problem in projecting the visual images is the computer's capacity. The more detail you show, the greater computer power it requires. In previous visual systems," he adds, "they were running close to the capacity of the computer. You just couldn't force any more detail. Any time you wanted to add another tree, it took up more of the computer's power."

To overcome this problem, CAE-Link has developed a system called ESPRIT, for Eye-Slaved Projected Raster InseT. A laser sensor follows the motion of the eyes. Wherever the pilot looks inside the dome, the image seems sharp and clear—but only where the pilot is focusing. The side-view areas, where the pilot would not detect sharp images anyway, are blurry. This saves computer power for the all-important area on which the pilot is concentrating.

For a pilot flying fifty feet off the deck at five hundred miles per hour, one mistake can be fatal. Simulators provide hundreds of hours of training before the real thing.
CAE-Link

A sensor mounted in the ESPRIT helmet tracks every movement of the eyes. The projector sends high-resolution pictures of whatever the pilot focuses on. The side-view, or peripheral, images don't need to be as clear.
CAE-Link

A system by CAE Electronics does away with the dome altogether. In the FOHMD, the Fiber-Optic Helmet Mounted Display system, images are projected inside the pilot's helmet. Fiber optics are strands of glass or plastic, finer than hair-thin, woven into cables that transmit signals at the speed of light. In the FOHMD, these cables carry images to lenses in front of the pilot's eyes. The trainee can look around in a full circle, not only at two-dimensional images, but at stereo-optic 3-D objects with height and width and depth.

This state-of-the-art CAE helmet works without a dome or a screen. Real-time images are projected right onto the lenses in front of the pilot's eyes.
CAE-Link

No matter which system is used, flight simulators don't come cheap, but usually they cost less than the real aircraft. An F/A-18 fighter jet costs $40 million; an F/A-18 simulator costs $20 million. An EA-6B aircraft has so much advanced electronic equipment inside it that the aircraft sells for $60 million. In the EA-6B simulator, all the fancy electronic gear is simulated by just one computer, which keeps the price down to $16 million. Also, simulators can work around the clock, and they don't need fuel. But their greatest economy is in the accidents they prevent.

The most advanced flight simulation system, soon to be operational, will use photographs gathered by satellites and reconnaissance airplanes. Instead of a computer programmer creating images that *look like* the real world, the system will use *actual photos* of the real world. They'll be digitized (converted into a format the computer can use), processed, and will emerge as high-resolution simulations—all within forty-eight hours! This will allow military pilots to rehearse real missions on the simulator shortly before takeoff. These accurate, up-to-the-minute simulations of mission objectives will increase pilot safety.

When the newest flight-simulation system becomes operational, military pilots will rehearse their missions only hours before flying them. The simulations will be made from real-world photographs gathered by satellites.
Evans and Sutherland

Through the windows of the Link Shuttle Mission Simulator, astronauts watch high-resolution images of scenes they'll see in space.
Evans and Sutherland

Computer-image simulators train not only aircraft pilots, but submarine personnel, tank drivers, astronauts, ship-deck pilots, and military personnel. Many of the high-tech weapons that were used in the Persian Gulf war are so expensive—$50,000 for a single laser-guided tank-busting shell, for instance—that troops were not able to train with real weapons. Instead, they practiced on computer simulations. Some low-tech, inexpensive desktop simulators can be taken right onto the training field for combat rehearsal, where missiles that cost tens of thousands of dollars to launch can be "fired"—on a simulator—for a few cents' worth of electricity.

Astronauts receive much of their pre-flight training in a simulator. Instructors can program 6,800 different malfunctions into the SMS—the Shuttle Mission Simulator—in Houston, Texas. Sometimes they throw in three or four different malfunctions all at once.

Pilot Mike Coats, who has flown two space missions, says, "The simulator takes a little getting used to. You have to learn to put up with the motion, and be strapped in and tied down the same time you're learning to reach for switches. After a few hundred hours in the simulator, though, you're fairly comfortable. And when you do climb into the real spacecraft, you feel right at home." Nearly every astronaut who's been in space has said that the real space voyage was exactly like the simulator.

You've probably played a video game in an arcade or on a home computer where you drove a car or flew a plane. With every move you made with the joystick or on the keys, the image changed instantly. You had to keep the car from crashing into a barricade, or keep the plane from going into a spin. You may not even have known the word *simulate*, but that's what you were doing—simulating flight or simulating driving. Says Neal Mayer, manager of data base engineering at Evans and Sutherland, "A video game is basically like one of our simulators, but much simpler, with very low-resolution graphics, because very simple programs are running it."

Add high-tech sound, motion, and visual images, and you get something as sophisticated as Disney's Star Tours ride, the world's best-known simulator (see frontispiece). As the Starspeeder 3000 climbs, banks, and accelerates on its way to the Moon of Endor, only to take a wrong turn toward Darth Vader's Death Star, 27,000 passengers a day experience the fun and excitement of a simulated space voyage.

Technically, the Star Tours rides are very much like flight simulators, the Shuttle Mission Simulator, or any of the other simulators that train pilots, astronauts, and navigators. The big difference is that while the rides are strictly for fun, trainer simulators serve the very serious purpose of saving lives.

With twelve missions and five skill levels, the Falcon F-16 Fighter Simulation game is so realistic, it has been selected as the basis for a military simulator.

Drive through San Francisco over the Golden Gate Bridge in the VETTE! driver-simulation game.

Author photos. Screen images by Spectrum HoloByte

The Cray Y-MP supercomputer, reflected in a shiny floor, can do 2.2 billion calculations per second. These are not normal operating speeds—supercomputers achieve peak speeds for only fractions of a second. But in one minute, this Cray can solve a problem that would take 96 hours on a home computer.

Photograph by Paul Shambroom, courtesy of Cray Research, Inc.

SEEING IS PERCEIVING *Chapter 5*

When the television news shows a space probe flying past a distant planet, what you're really watching is a simulation of the planetary fly-by. The actual flight past the planet can't be telecast because it's happening millions of miles away, too far for the signals to reach us. You'll usually see the words "This is a simulation" running across the bottom of your TV screen. Computers have taken all the data about the space voyage and turned the information into clear pictures for you to watch.

Supercomputers can simulate almost anything in the known universe today. All physical information that can be written as formulas or numbers can be turned into computer images.

Before computers were invented, physicists and mathematicians filled reams of paper with equations about the physical world. Today, computers can take these pages and pages of calculations and turn them into pictures. Bob Judd, a computer expert at Los Alamos National Laboratory in New Mexico, says, "When you have a machine calculating a billion numbers

Twelve trillion calculations for each image made the tin toy more vibrant than life. Five years from now, personal computers should be able to produce images as brilliant as this.
© 1988 Pixar

per second, and you run it for ten hours, think of all the calculations that have happened in that span of time. It would be very hard for a human being to sit down and understand the results from all those numbers. So you try to break it down to something you can relate to. Computer pictures are things that people can understand."

At the computer terminal, a software program instructs the computer to build with red, green, and blue beams of electrons that make *pixels*—picture elements. The more pixels on a screen, the sharper the image will be. Low-resolution home computer screens start at about 64,000 pixels; the best television screens hold about a quarter of a million pixels; and standard high-resolution computer screens now have about 1.3 million pixels.

The computer is instructed—either by the software program or by a scientist or artist wielding an electronic pen, a joystick, or another interactive device—about what shapes are needed. It's told how the light hits those shapes, how they look from different angles, and how they move. Sixty times a second, the computer redraws the picture, updating its position. This takes billions of calculations per second.

Computers need huge amounts of memory to store all the information about position, color, shape, and design in each picture. The higher quality the picture is, the more instructions are required to create it. The dazzling clarity that won an Oscar for the animated theatrical movie *Tin Toy* required 12 trillion computer calculations per frame.

Scientists also use the term *movie*, but they don't mean the kind that comes out of Hollywood. Their movies are created with mathematical equations and are played on computer screens. These "movies" are also called scientific visualizations. They let scientists discover things they might not otherwise see.

Chuck Hansen, a computer graphics specialist at Los Alamos, works on climate models dealing with global warming. Using a computer, he takes numbers listing the earth's temperatures and assigns a color to each number, ranging from red for hot to blue for cold with other colors in between. On computer or video screens, the electronic beams that create pixels can combine red, green, and blue into sixteen million different colors.

Next Chuck Hansen "wraps" the color-coded temperature pictures around a three-dimensional computer image of the earth's sphere. On the screen, the sphere revolves. Hansen says, "The nice thing about this simulation is that the scientists can say, 'Well, I want to see what's going on in the Himalayas, or at the South Pole.' So we grab onto the image and tilt it. This

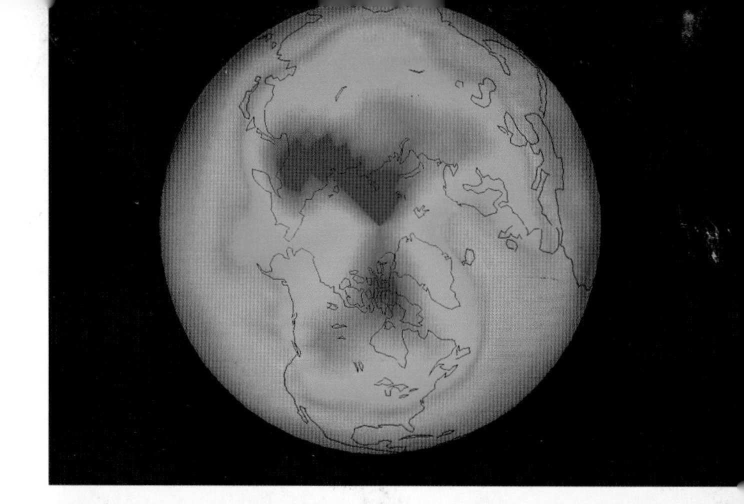

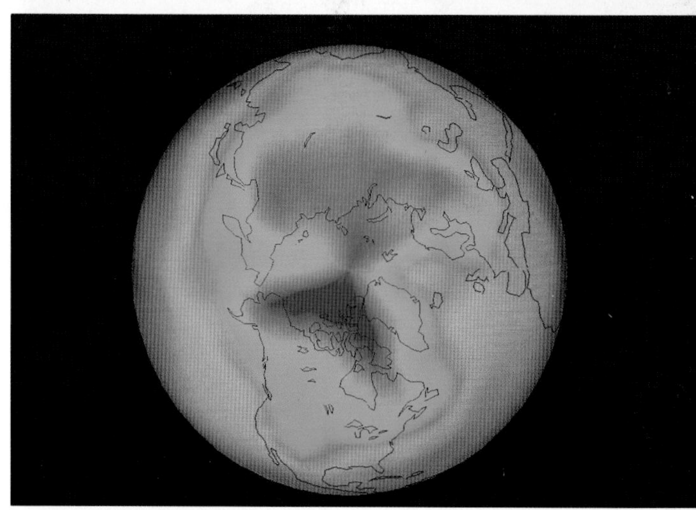

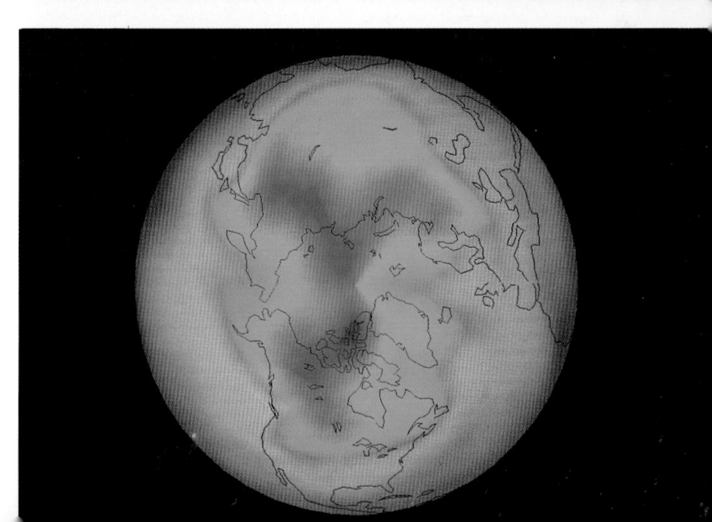

A computer movie showing the earth's changing temperatures can be speeded up and tipped in any direction. Suddenly, patterns of warming and cooling become visible.
Chuck Hansen, Los Alamos National Laboratory

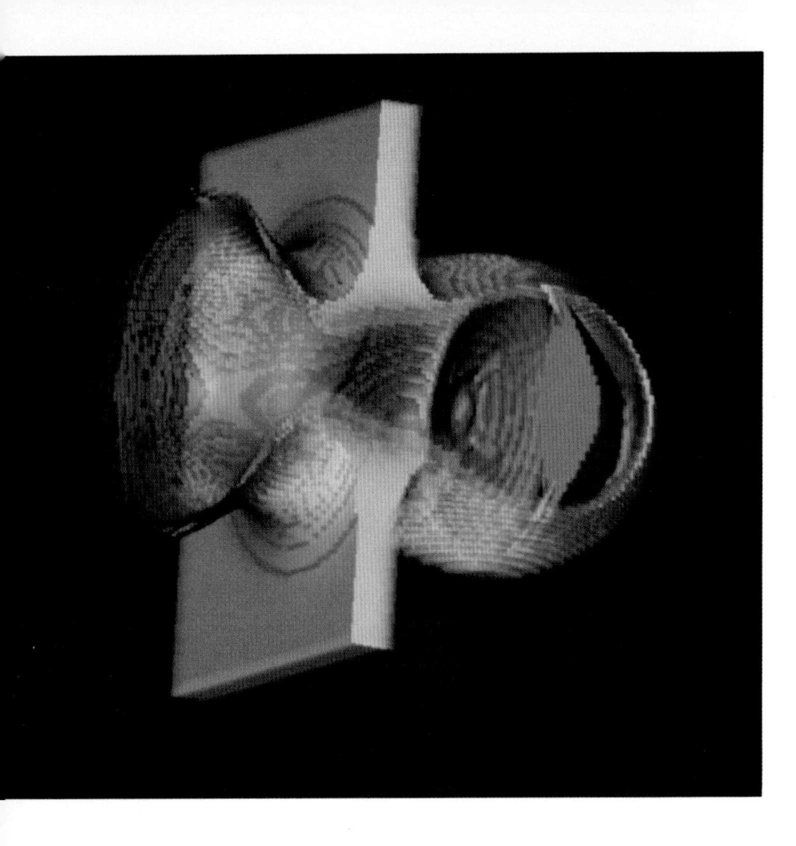

Images can be stopped for closer study. Here a lead ball penetrates a lead plate in a computer simulation. Color coding shows differences in the metal's density.
John Fowler, Los Alamos National Laboratory

is a good tool, because rather than looking at sheets of paper, we can look at a movie and play it back to see how things progress." In the speeded-up version, six months' worth of temperature data is presented at the rate of four days per second. Viewed at this rate, the total six-month temperature sequence takes only forty-five seconds. Scientists study the sequence over and over again to become aware of subtle changes in the simulation. They notice warming patterns that would be missed by someone looking at endless rows of numbers.

Computer simulations can be slowed down, too, to tick away a frame at a time. Or the scientist can stop them to ponder a single step in a process.

Computer models also show how one part of a model relates to the other parts. In a model showing molecules in a crystal, for instance, you can tell which molecule is which because the computer has color-coded them. This makes identification easier than it would be under even the highest-power microscope.

Today, computers can calculate the way air flows over a wing shape and turn the numbers into pictures on the screen. When the design of the airfoil is changed, the computer images show the changes in wind speed, stress, and temperature. The computer is programmed to analyze each change until the simulation appears error free. Only then do engineers build a physical model for wind-tunnel testing.

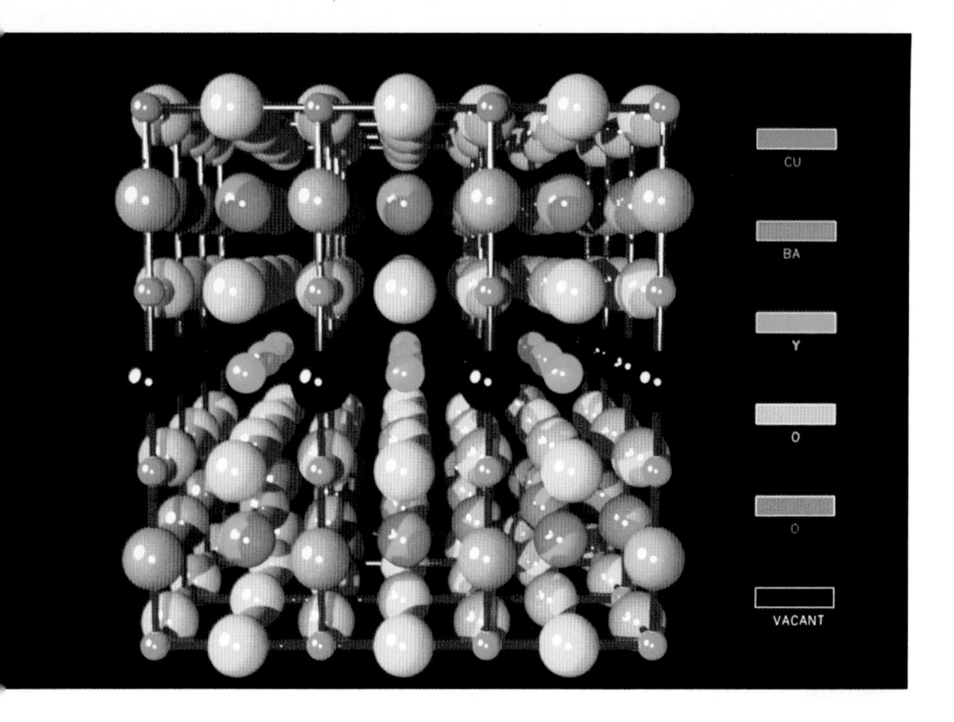

Different colors clearly show the structure in a model of a high-temperature superconductor crystal.
Melvin L. Prueitt, Los Alamos National Laboratory

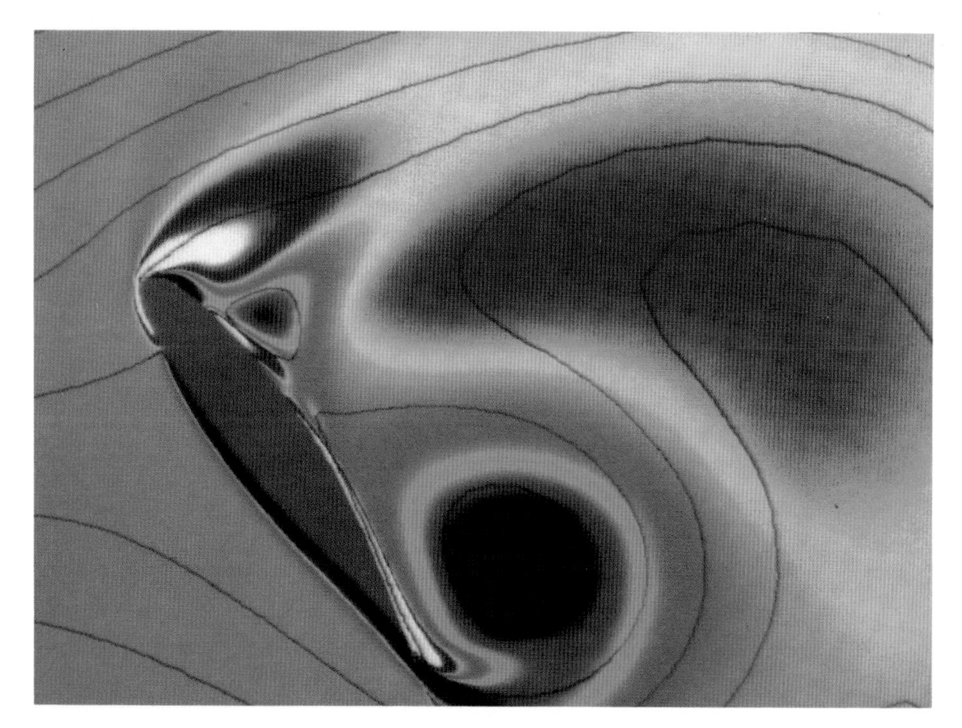

A computer image of air flowing over a wing not only lets engineers measure airflow, it allows them to change the shape of the airfoil by programming different numbers into the computer. Then they don't have to build a new physical model for each change.
NASA Photo

Air pollution and earthquakes—physical simulations you read about earlier in this book—can also be simulated with computer images. Computer movies show how pollution from a single large factory can spread across the continent. Scientists studying earthquakes can calculate the forces on a single grain of sand, or zoom onto the image of a planet and observe how earthquakes shaped it.

By speeding or slowing time, by magnifying, sorting, coloring, tracking, and picturing everything from molecules to galaxies, computer simulations show people patterns they would otherwise miss.

A section of spiral bacteria would never be this easy to see under an electron microscope.
Melvin L. Prueitt, Los Alamos National Laboratory

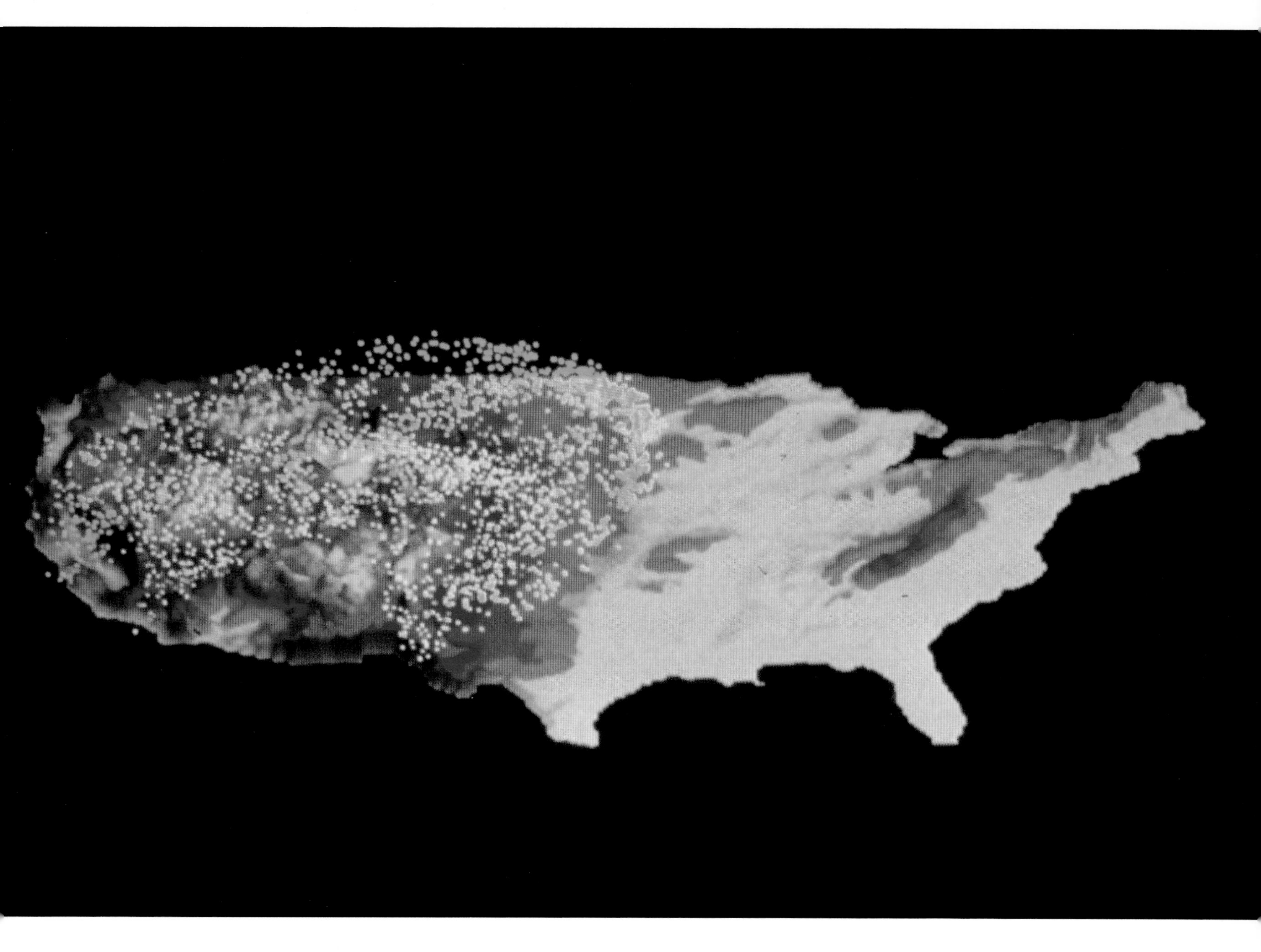

In a nuclear winter study, particles are traced across the United States from a point of origin on the California coast.
Melvin L. Prueitt, Los Alamos National Laboratory

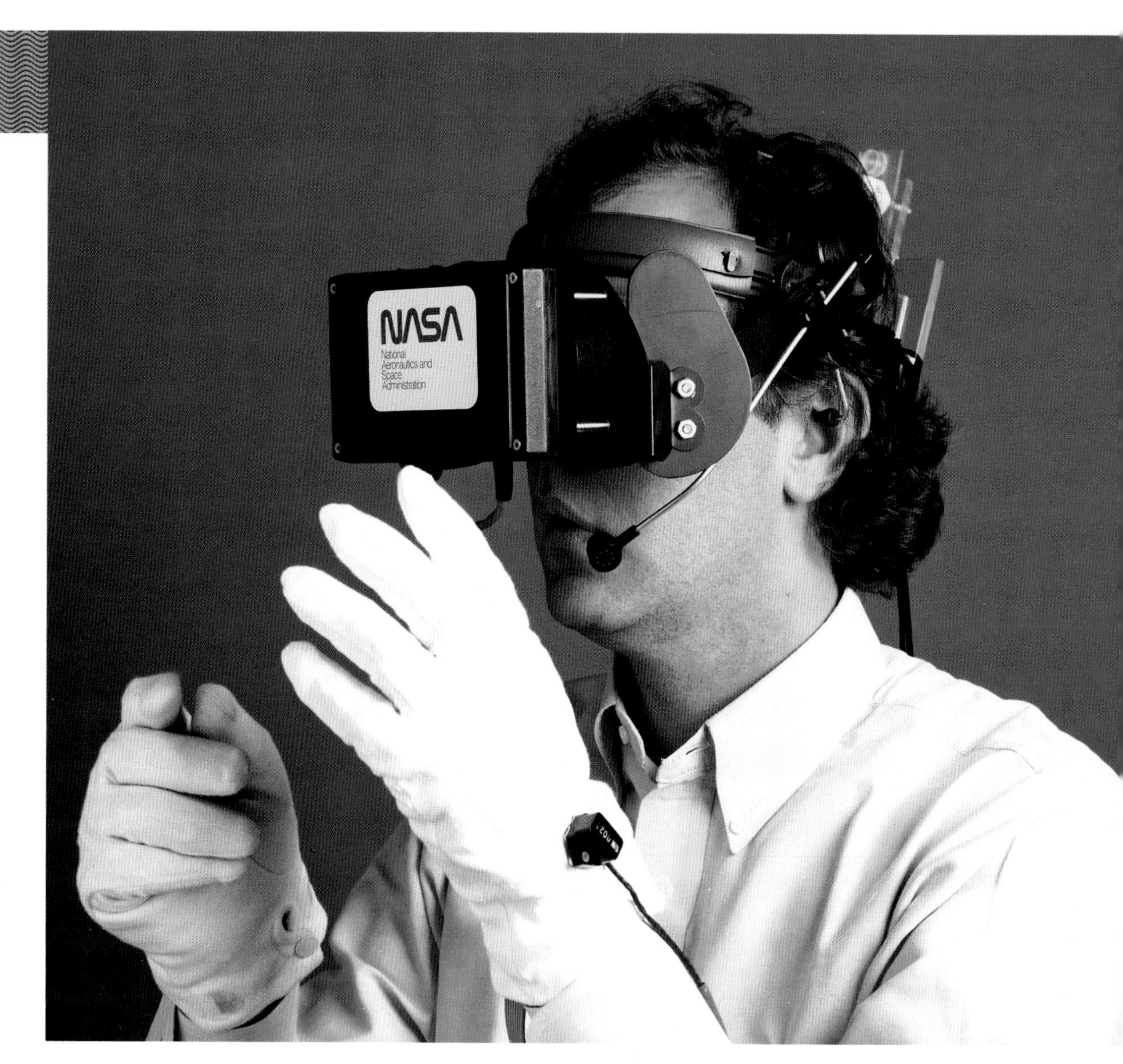

Scott Fisher, wearing a virtual reality headset and gloves, says, "The bottom line here is to go from personal computer to personal simulator, to be able to simulate whatever it is you'd like to move around inside of."
NASA Photo

Link's FOHMD, the helmet-mounted display system described on page 42, lets wearers turn their heads to see everything around them. They see computer-generated images, mostly to do with flying, since so far the helmets have been used only for flight training. The same kind of helmet is being developed at NASA-Ames Research Center, not to look at computer simulations, but to see what a robot in space will see.

Since it has television cameras in place of eyes, the space-walking robot can focus on real objects, and the pictures will be telecast to two small television screens inside an astronaut's helmet. Through a process called teleoperation, the astronaut can then move the robot's arm—to repair a part of the space station, or to do other work outside the spacecraft. What the robot looks at, the astronaut will see. When the astronaut moves, the robot will move in exactly the same way.

Now imagine that instead of the images coming from robot "eyes," they're coming straight from a video camera into goggles

A DataGlove wearer doesn't see the screen, just the images projected on the inside of his goggles. He can reach right into the picture to touch any part of it.
VPL Research, Inc.

you're wearing. Everywhere you look, the camera moves instantly to look at the same spot. No matter where you turn—above you, all around you—you see only what the camera sees, because the goggles block the real world from your view. If you hold your hand in front of your face, you can see your hand only as the camera sees it.

Take it a step further. The images that you're looking at are not of the real world. They may be entirely imaginary, like dragonflies with alien soldiers riding on their backs, or dancing flowers, or golden coins and chains falling out of silver clouds. Not only can you see these things, you can reach out to catch the gold coins or to pluck one of the dancing flowers—and watch your hand go right inside the picture with them. You can clench your fist to catch the alien or cup your palm to lift the dragonfly. You can *become* the dragonfly, so that when your arm moves, the image's wing moves exactly the same way.

What you're seeing is not some science-fiction fantasy of the future. This future is already here, and it has a name: virtual reality. It has three names, actually, because the technology is so new that no one has yet decided exactly what it should be called—virtual reality, artificial reality, or virtual environment. In computer terms, *virtual* means something that exists only as a computer-generated image.

In the offices and lab of VPL Research, Inc., in Redwood City, California, some of the most creative people in all of technology are producing virtual environments. They've built a glove called DataGlove, lined with fiber optics. When held straight out, it sends beams of light over cables for a computer to process. When the wearer's hand bends, the light beams are interrupted, and the signals received by the computer change.

The computer notices the changes, calculates the new, adjusted images, and sends them back to the Eyephone, as the eye-tracking system inside the goggles is called. The images are simple—almost cartoon-like—and a bit fuzzy because they require so much computer power. Computers aren't yet advanced enough to make picture-perfect images. But even if the pictures aren't altogether sharp, wearers have no trouble believing in the things they see. "There is an amazing phenomenon in virtual reality," says VPL's founder, Jaron Lanier. "It works better than it should because your brain wants your environments to look good. [Your imagination can] cover up gaps in our simulations." In other words, your mind just slips over the parts that aren't exactly right. In a similar way, when you play a video game, you throw yourself wholeheartedly into the action as though it's real, even though the pictures on the screen don't look like the real world.

Fiber-optic cables on the DataGlove use light to communicate with computers. Bending the fingers interrupts the flow of light, changing the signal.
VPL Research, Inc.

What is all this and what is it for? At NASA-Ames, virtual reality's main use will be with the teleoperated robot. That's a practical use, since robots are safer and cheaper than human astronauts for performing dangerous repair jobs outside a spacecraft.

But aliens on dragonflies?

VPL's Jaron Lanier, who creates most of the computer images for virtual reality, says, "It's a reality in which anything can be possible. . . . It's a world as unlimited as dreams."

At the start of the nineties, hardly more than two thousand people in the whole world had ever experienced virtual reality, because it's in limited supply and it's very expensive. A DataGlove costs $8,800; the Eyephones are $9,400; and the computer graphic workstations powerful enough to run the system cost another $49,900. That's the price of equipment for

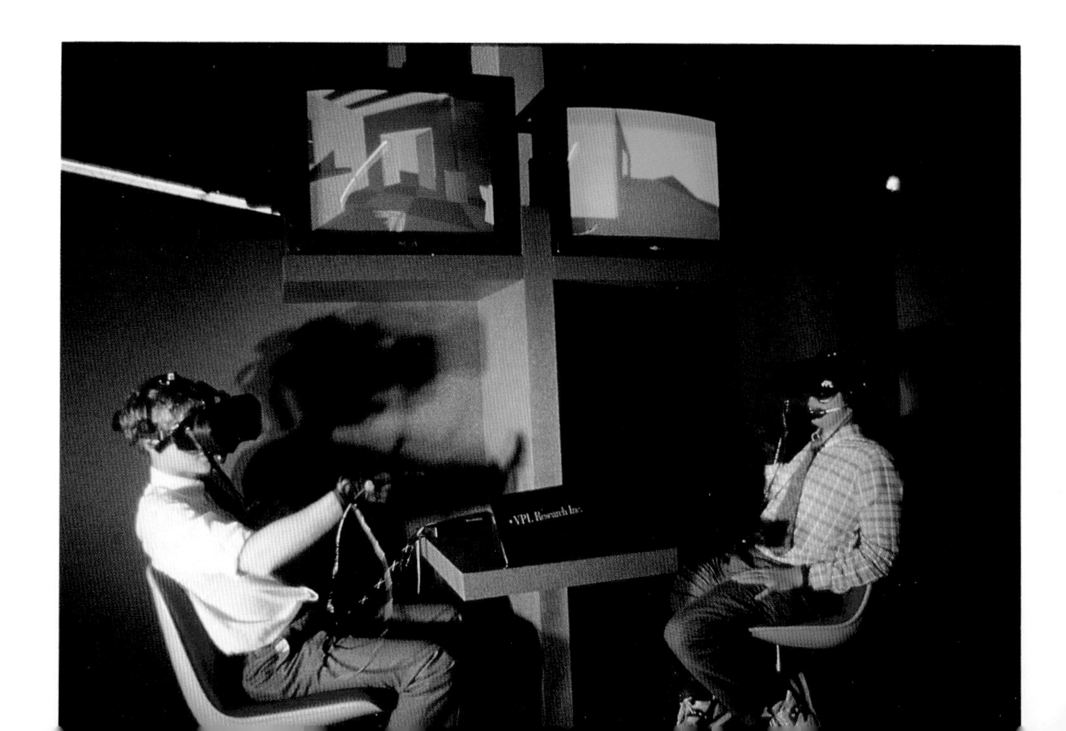

VPL's founder, Jaron Lanier, believes that virtual reality works best as a shared experience. Each of these users can enter the other's virtual world.
VPL Research, Inc.

just one person, and Jaron Lanier believes that virtual reality is meant to be shared.

For now, though, such sharing is financially out of reach. A virtual reality system that two people can use together—each one in a DataSuit that allows the entire body to enter the picture (see page 61)—costs close to a half-million dollars.

A basic, scaled-down, low-tech version of the DataGlove is available right now in toy stores for about $80, not $8,800. VPL Research, with Abrams/Gentile Entertainment, Inc., has licensed this system to Mattel Toys for use in Mattel's Power Glove; it works with certain Nintendo video games. When you bend your fingers in the glove, you send a signal to the black box on the TV set, and Mario jumps. So far, the Power Glove is mostly a substitute for a joystick. But newer games, such as Super Glove Ball, will react slightly more like virtual reality.

NASA predicts that when advanced virtual reality becomes affordable in the future (high-tech devices almost always come down in cost over time), schools will own it, letting students "explore the Colosseum or the streets of Rome, or enter the arteries and veins of the heart" in simulated images. Even in the present version, virtual reality has practical uses. Architects walk through floor plans with it—if a virtual room feels too cramped to them, they can rearrange a wall just by pushing it around in the computer image. With a DataGlove, scientists can actually reach a hand into a computer model of a molecule and change the position of an atom. Wearing DataSuits, people walk around inside models of communication networks, or traffic-

Mattel's Power Glove is a low-tech spin-off from the VPL DataGlove. With it you can control several Nintendo games, including Super Mario Bros. 3.

Author photo. Power Glove is a trademark of Mattel, Inc. Used with permission. Super Mario Bros. 3® used with permission of Nintendo of America, Inc.

The earliest simulations? During their rituals, prehistoric people interacted with images.

© *National Geographic Society*

control systems, to locate tie-ups. By adjusting an image, they can change what happens in the real world.

Fifteen thousand years ago, in dark caves located on continents far apart from each other, prehistoric people experienced trances and saw visions. In these trances, they believed they entered into a spirit world where they actually became one with the animals they would hunt—talking with them, taking on their appearances, thanking them for food for the tribe. Today, anthropologists such as David Lewis-Williams, professor of archeology at the University of Witwatersrand, Johannesburg, South Africa, believe that the stone-age cave paintings are not just pictures of everyday life in prehistoric times, but images of the mystical world these people entered during their trances. What they did in those visions—asking the spirits for good hunting, seeking to control the movements of the game animals, bargaining for good weather—would change the real world they lived in, they believed.

These prehistoric people had no technology. They were trying to understand, interact with, and change their environments in the only way they could, using their minds.

In the twenty-first century, human beings will use powerful computers to reach worlds of our own imagining, our virtual realities. With advanced technology, perhaps we will enter our simulations and find the answers to difficult questions—touch our creations, change them, improve them. Move backward and forward with them over time and space to see where they fit. Create virtual worlds that will show us, in simulations, solutions to problems we can't even guess at now.

Director of product design at VPL Research Ann Lasko-Harvill interacts with a virtual reality she designed for herself. She says, "People researching in these fields always work on the very edges of what the technology can support. There's an excitement and a readiness for the next step."
VPL Research, Inc.

Glossary

aerodynamics the study of air flow; the science of determining how the shapes of objects are affected when air or other gases move over or around them

airfoil a shape, especially of a wing, designed to give aerodynamic advantage to an aircraft in flight

artificial reality video simulations that produce the illusion that the user is inside a computer image

base isolators flexing devices placed between the foundation of a building and the building itself to reduce the force of earthquake shocks on the building

computer-generated image an image drawn by a computer based on data from pictures or from human input

computer modeling using the power of a computer and specialized software to generate images of objects, terrain, or scientific phenomena

computer simulation a representation of a real event or an object, created from a computer program

data a collection of observations and facts, often generated and processed by a computer

DataGlove a glove lined with fiber optics that lets hand signals operate computer programs in virtual reality

digitize to convert photos or drawings into values that can be processed by a computer

Earth's gravitational field the space around the Earth in which an object will be pulled toward the center of the Earth

Eyephone a head-mounted system with stereo sound and visual display screens for three-dimensional viewing, used in virtual reality

fiber optics tiny threads of glass or plastic that transmit data on light waves

high-resolution graphics video or computer-screen images produced by a large number of pixels. The greater the number of pixels, the clearer and sharper the image will be

hydraulic operated by a liquid under pressure

hypersonic extremely high supersonic speeds above Mach 4

input data entered into a computer program by an operator using a keyboard, joystick, or other device

interactive a system in which a user provides input and receives response from a computer program. Communication between the computer system and the user is almost instantaneous.

low-resolution graphics video or computer images produced by a small number of pixels. These images do not have the clarity and sharpness of high-resolution graphics.

mock-up a physical model, usually full-scale

model a representation of a system or an object to be studied. It can be a physical copy of any size, or a computer-generated image.

neutral buoyancy to neither sink nor float, but remain at a constant depth underwater

orbital speed above 18,000 miles an hour, fast enough to break free of Earth's gravity and circle the Earth

parabola a curve generated by a specific mathematical equation, such as the pattern of a falling object responding to the force of gravity

physical simulation one that can be touched or felt, that has the appearance or behavior of a real object

pixel picture element; the smallest dot of light on a computer screen or television screen

real-time visuals computer-generated images that move at the same speed that events would occur in the real world

Richter scale a method of measuring the magnitude of earthquakes

scaled-down smaller in size, but identical in shape, to the real object. "Scaled to one-third" means one-third of actual size or weight.

scientific visualization computer graphics that represent a scientific concept or object

shock tunnel a tube in which short bursts of gas under high pressure flow over a model, simulating hypersonic speeds

simulation an imitation that represents a real object, like an airplane's cockpit; or represents a force, like the wind; or an abstract idea, such as nuclear winter

software program a program that controls a computer

space probe a satellite or other spacecraft sent to gather data in outer space

speed of sound about 760 miles per hour at sea level. Also called Mach 1.

supercomputers the biggest, fastest, most powerful computers made, able to do billions of calculations per second

supersonic faster than the speed of sound, between Mach 1 and Mach 4

teleoperation the manipulation of a robot or other piece of equipment by an operator some distance away from it

turbulence violent disturbances in airflow

validation a proof of the accuracy of data by comparing it to a real object or occurrence

virtual appearing to be, rather than actually being. Or, something that exists only as a computer image.

virtual reality same as artificial reality; also called virtual environment

visual systems computer-generated images that look like real-world scenes in flight simulators or other simulators used in training programs

zero gravity a balance achieved as the gravitational pull of Earth is counteracted by other forces, such as centrifugal force, when an object is in orbit. Weightlessness.

Index